ENDANGERED SPECIES

BY
Caroline Evensen Lazo

CRESTWOOD HOUSE

New York
Collier Macmillan Canada
Toronto

Maxwell Macmillan International Publishing Group
New York Oxford Singapore Sydney

Library of Congress Cataloging-in-Publication Data
Lazo, Caroline Evensen
 Endangered species / by Caroline Evensen Lazo. — 1st ed.
     p.   cm. — (Earth alert)
 Includes bibliographical references.
 Summary: Describes the plight of a variety of plants and animals threatened with extinction and what can
be done to save them.
 1. Endangered species—Juvenile literature. 2. Rare animals—Juvenile literature. 3. Rare plants—Juvenile
literature. 4. Wildlife conservation—Juvenile literature. [1. Rare animals. 2. Rare plants. 3. Wildlife
conservation.] I. Title. II. Series.
 QL81.L39   1990   574.5'29—dc20   90-35494   CIP
 ISBN 0-89686-545-2   AC

Photo Credits
Cover: Peter Arnold, Inc.: (Kevin Schafer)
Animals Animals/Earth Scenes: (Stouffer Productions) 4; (Bruce Davidson) 9; (Jack Wilburn) 12; (G.I.Bernard)
    17; (D.H. Thompson) 18; (J.H. Robinson) 21; (L.L.T. Rhodes) 25; (Johnny Johnson) 30; (Richard Packwood/
    Oxford Scientific Films) 33; (Patti Murray) 35; (Richard Kolar) 39; (Michael Dick) 40
Devaney Stock Photos: (B.J. Adams) 7; 14, 23
Peter Arnold, Inc.: (Norbert Wu) 26
AP—Wide World Photos: 29
Four by Five, Inc.: 36-37

Macmillan Publishing Company   Collier Macmillan Canada, Inc.
866 Third Avenue   1200 Eglinton Avenue East
New York, NY 10022   Suite 200
                                  Don Mills, Ontario M3C 3N1

CRESTWOOD HOUSE

Produced by Flying Fish Studio Incorporated

Printed in the United States of America

10 9 8 7 6 5 4 3 2

CONTENTS

INTRODUCTION

While hiking in Yellowstone National Park, a group of Girl Scouts was warned about the grizzly bears that lived there. The campers were nervous and a little scared—until they saw a sign stuck to a tree. BEWARE OF MAN, it read. And someone had signed it, FATHER BEAR. Everyone laughed, and they continued their hike.

Later a ranger explained why the sign was no joke. "Many people forget that this is the home—the habitat—of the grizzly bear and other wild animals," he said. "We must respect their right to live in peace. They were here long before you and I were."

He reminded the campers that grizzly bears will not attack unless they are tempted to. They have a superior sense of hearing and smell. They can't resist the scent of bacon grease, for example, and will follow it anywhere. "Keep food covered," he warned. "The more you know about the bears, the more you will respect them."

For many years, humans killed bears for their fur and cut down their forests to make roads and houses. Today there are only one thousand grizzlies left in the United States south of Alaska. To

Grizzly bears at play. Hunters and careless campers have destroyed so much of the bears' habitat that they have been put on the endangered species list.

5

prevent further loss, the grizzly has been placed on the endangered species list. This means that the bear is protected by federal law. To harm a grizzly is a crime. Those who damage the bear's habitat will be punished, too. But can laws alone change the bad habits of humans?

Naturalist Aldo Leopold said it would take laws, education, and compassion to save our wildlife. Each of us must make a promise to do what we can to save all the threatened species—from the tiny .062-ounce bat to the 6.5-ton elephant. But, in order to keep our promise, "we must first see, feel, understand, love, or have faith in" all species, Leopold said. We are all connected in what he called the "web of life." All living forms—including plants—are interdependent. Losing just a few can endanger humans, too.

SHARING A GOAL

When Delia and Mark Owens left their comfortable home in the United States to live among the lions in southern Africa, their friends were not surprised. Delia and Mark had become friends while studying zoology at the University of Georgia. They shared the same goal: to learn as much as possible about wildlife and to help save threatened animals from destruction. When they heard that two-thirds of Africa's wildlife had been destroyed, they were alarmed.

When they arrived in Africa, Delia and Mark were shocked at what they saw. They saw ancient animal habitats being cleared to

Delia and Mark Owens traveled to the Kalahari to study animals like this African lion. Their findings led to new protection laws being passed.

make room for farms and roads. They saw animals being shot, trapped, fenced in, and poisoned to aid people's business and pleasure. Delia and Mark hoped that their study of the lions, hyenas, and other creatures of the Kalahari Desert would lead to world action to save the wildlife there.

Many African countries work very hard to protect wildlife. They have large parks and strong laws against poaching. But many African countries are also poor. They do not have enough money to maintain parks or to pay park rangers.

Money is less of a problem in the United States. In spite of our national wealth, our wildlife protection record is not good. Even here, poachers threaten wildlife. And many species of animals have become extinct in this country.

Rich Block of the World Wildlife Fund says, "The biggest thing we are short on is compassion." To save animal species, he says, we need to remember that we are all in one world. We need to work together. To do so, we must learn about the past. We must understand how humans became wildlife's most feared predator.

IN THE BEGINNING

More than three billion years before the age of dinosaurs, many other creatures lived on our earth. Some became extinct through evolution, the gradual process of growth and change.

African countries find it very difficult to control poaching. These leopard skins were confiscated by officials in Kenya.

Below is a list of some of the animals and plants currently classified as endangered species. The list comes from the U.S. Fish and Wildlife Service of the Department of the Interior. It is always changing, with some species being added and some being taken off as their situations improve. The bald eagle, for example, has recently been reclassified as a threatened species because of successful attempts to increase its number. This means that technically the species is no longer endangered.

Mammals

African & Asian Elephant
Bactrian Camel
Beaver
Black Rhinoceros
Blue Whale
Bobcat
Brown Bear
Brown Hyena
Caribbean Monk Seal
Cheetah
Chimpanzee
Dugong
Ferret
Florida Squirrel
Giant Armadillo
Giant Otter
Giant Panda
Gorilla
Gray Bat

Gray Wolf
Grizzly Bear
Howler Monkey
Jaguar
Lemur
Leopard
Lion
Manatee
Orangutan
Porcupine
Red Kangaroo
Sloth
Spider Monkey
Tapir
Tiger
Wild Goat
Yak
Zebra

Birds

Australian Parrot
Brown Pelican
California Condor
Florida Scrub Jay
Golden Parakeet
Hawaiian Hawk
Ivory-Billed Woodpecker
Peregrine Falcon
West African Ostrich
Whooping Crane

Reptiles & Amphibians

Boa Constrictor
Chinese Alligator
Crocodile
Iguana
Rattlesnake
Salamander
Tortoise

Fishes

Catfish
Chub
Sturgeon
Trout

Plants

Arizona Hedgehog Cactus
Cook's Holly
Dwarf Lake Iris
Elfin Tree Fern
Green Pitcher Plant
Lakeside Daisy
Northern Wild Monkshood
Scrub Mint
Spurge
Swamp Pink
Truckee Barberry

Illegally trapped bobcats and coyotes are a grim reminder of wildlife's worst enemy—greedy hunters.

Some were killed by earthquakes, floods, and volcanic forces. Changes in climate killed others. Many species died out as smaller, faster ones took their food and space. The most common threat to them came from predators—animals who prey on others. The ones that could outfox their predators and adapt to the climate were apt to survive.

When climate changes slowly, animal species have much time to adapt. This process is called mutation. "New shoots sprouted from the best of the old," said naturalist James Cox, "and the quality of organic life became enriched."

Gradually forests replaced deserts, and plants began to grow along riverbanks. Fresh water, trees, and plants provided the habitats for new species of mammals and birds that began to form our natural world. This close kinship between the land, sea, and all the species is called the ecosystem. Even the smallest moth plays a big role in it.

Aldo Leopold compared the smaller species to cogs in a wheel. They keep the wheel moving. Moths, for example, pollinate the rain forests and provide food for birds. Since all species are interdependent, when one dies out, another is sure to follow. It is crucial to save all species. If we do not, our own species could become extinct.

We no longer need the bear's fur to keep us warm or the whale's oil to light our lamps. Yet whales are being killed every day. Some people wonder if the human "hunting instinct" is too strong to ever be overcome. Psychologists say that human desires for community, peace, and love are instinctive, too. With effort, these values can transcend greed and selfishness.

Harp seals are in danger of becoming extinct, like the Caribbean Monk Seal, due to overhunting.

SEEDS OF DESTRUCTION

When Christopher Columbus discovered the New World, he also discovered the Caribbean Monk Seal, as noted in his journal. Later hunters found the mammal, too. They killed it for its skin and oil. By 1952 the Caribbean Monk Seal had vanished forever; it became extinct.

The first wanton killing of animals dates back at least to 1680 when Dutch settlers clubbed the dodo to death for fun and sport. The moa, a flightless bird native to New Zealand, was hunted for its feathers and became extinct by 1800. In the 1850s, passenger pigeons made up almost half of North America's bird population. Then hunters went after them for fun and sport, killing more than one million a year. Now passenger pigeons are also extinct.

Plants, too, have suffered from the selfish acts of humans. According to *Science* magazine, 680 plant species now living in the United States will become extinct between 1990 and the year 2000. That figure is three times greater than the number of plant species that died out during the entire two centuries before!

Elephants are one example of an endangered species. In the past ten years, elephants have been killed at a shocking rate — one every ten minutes. Poachers search out elephants and kill them. They tear out the ivory tusks and leave the carcasses to rot.

In 1989 a two-year, worldwide ban on the trade of ivory was declared by CITES. This stands for Convention on International Trade in Endangered Species.

If no one buys or sells ivory, there will be no reason to kill elephants for their tusks. But as long as a market for ivory exists, poachers will continue to kill elephants. Some poachers are even hiding the ivory, waiting until the trade ban is lifted.

Some countries do not agree with the ban. In Zimbabwe, for example, elephants have been protected. Zimbabwe's elephant population is actually growing. Why, people ask, shouldn't they be able to sell tusks from old or sick elephants who die? After all, they need to make money to continue protecting wildlife. But others say that when elephants die, they should be allowed to rest in peace. And to sell their ivory tusks—even for a good cause— will keep the ivory trade going.

ROAD TO DECLINE

Rain forests cover only about two percent of the earth. But they provide homes for over half of the world's species of plants and animals. Although there are rain forests in Asia and Africa, the largest area is in South America. The rain forest there is being cut down at a rate of 20 football fields a minute. The destruction of the Amazon rain forest is causing many environmentalists to worry.

Why? "Man needs roads ... and he needs more space to live," a farmer said. "There are more people here than ever before. Too many people!" Progress means that roads have to be built. It also means that land has to be turned into farms. But other things happen as well.

The lives of many plants and animals are upset when forests are cut down. The projects drive out people who live in the rain forests. The Yanomami people live quietly in Brazil. They collect

The demand for ivory has caused an alarming rise in the number of elephants killed by poachers.

Rain forests are home to over half the world's animal and plant species. Because of development, these vital habitats are being destroyed.

nuts and tap trees for sap. They make many medicines from plants. When forests are cut down, the medicines cannot be made. So the cures for diseases may be lost.

Rain forests are being cut and burned in many other places. Many scientists think that in a few years half of the world's rain forests will be gone.

In North America, less than 5 percent of the redwood forests are in their original state. Some are preserved in parks. However, many are privately owned and taken away by loggers every day. "I can't resist the money I can get for a redwood tree!" one owner said. But what happens to the 193 species of wildlife that use the redwoods for food, shelter, and other habitat needs?

Forty years ago Aldo Leopold urged mankind to treat all wildlife—including plants—as we would treat each other. Instead of acting like conquerors, ecologists say we should form a kinship with the rest of the natural world.

ROOTS OF CONSERVATION

Although America's laws to protect wildlife are the best in the world, they did not happen overnight. Laws were passed only after years of constant effort on the part of individuals and groups from New York to California.

In 1872 the United States government showed its concern for wildlife by naming Yellowstone the first national park. But it was

President Theodore (Teddy) Roosevelt who brought the subject into focus and into American homes.

While on a hunting trip, Teddy Roosevelt refused to shoot a grizzly bear. The event was talked about on the radio and printed in the newspapers. Americans began to think about the bear and its habitat. As a symbol of the event, a stuffed animal was made and named after the president—the teddy bear.

In 1903 Roosevelt signed an executive order to protect egrets, herons, and other birds on Florida's Pelican Island. The island became the first national wildlife refuge. But it was not until 1966 that the National Wildlife Refuge System Administration Act went into effect. The Refuge System is a network of lands and waters set aside for wildlife. The land provides habitat—food, water, shelter, and space—for about 60 endangered species. Hundreds of other birds, mammals, reptiles, fish, and plants are cared for, too. There are 400 refuges in the United States today.

In 1970 the EPA (Environmental Protection Agency) was formed to control pollution, a major cause of death among species of fish and wildlife. The Endangered Species Act followed in 1973. This act allowed for the greatest protection of threatened wildlife in the world.

In the mid-1970s, the Convention on International Trade in Endangered Species (CITES) was formed. Many nations joined CITES. Most believe the organization has done a good job overall. But the agency is abused by both business and government officers. Fake forms are filled out. Names are forged to get animals and products across borders. Local officials are often bribed by smugglers. To them, money means more than wildlife.

Poachers still get away with murder—the murder of the elephant for its ivory, the whale for its oil, the bird of paradise for

Clouds settle in a valley in Yellowstone National Park. In 1872, it became the first place set aside for wildlife by the U.S. government.

its plumage, and reptiles for their skins. Both legal and illegal trade of animals goes on. As long as men and women want to own fur coats, ivory jewelry, and snakeskin belts, more animals will die.

WHO IS IN CHARGE?

In the 1980s, the EPA was starved for funds and presidential leadership. Some companies became irresponsible. In one year (1988) 6,700 oil spills were recorded. The largest oil spill in United States history—10.5 million gallons—took place in 1989 off Alaska's coast. The oil tanker, *Exxon Valdez*, struck a reef in Prince William Sound. The ship's owners, Exxon, were slow to respond. A spokesman for the National Wildlife Federation said, "The oil industry has been dragged kicking and screaming into doing anything about oil-spill response."

An oil spill kills wildlife—from the small krill to the great whale. For example: Whales feed on the tiny shrimplike krill in Alaska's waters. The Exxon oil (and the chemicals used to clean it up) poisoned the krill. If the krill die out, will the whale die next? Already, the ecosystem is weakened.

"By weakening the ecosystem," says wildlife explorer Janine Benyus, "we sabotage its power to heal and nurture all life, including human life." Again, we are reminded of Aldo Leopold's "web of life." We are all linked together in the natural world.

An oil spill can be disastrous for wildlife. The sticky oil kills animals and poisons the water and food supplies.

In 40 years, the human population will double. Will this mean more highways for humans and less habitat for wildlife? Or will ten billion people at last learn to share the earth with the rest of the species that also live there?

THE MEDIA AND THE MESSAGE

Like Teddy Roosevelt, President John F. Kennedy aroused America's concern for wildlife—and for the human habitat as well. But unlike Roosevelt, President Kennedy was not a hunter. He was an ardent reader. In a major television speech, he urged all Americans to read *Silent Spring* by Rachel Carson. In her book, she warned about the dangers of insecticides and their fatal effects on wildlife and man. "If we don't take care of our world, we will lose it," she said. That book was written almost 30 years ago.

With the help of TV and the printed media (newspapers, magazines, and books), we can see the state of wildlife in much of the world. Reports range from *National Geographic* specials to local news spots. In 1989 people in Nevada cheered when they watched the fate of the desert tortoise unfold on the TV screen. They saw a multimillion-dollar building project come to a halt in order to save the endangered turtle from extinction. "A few years ago no one would have cared," a news reporter said. "We've come

All species—including ours—can become endangered if people continue to pollute their environment.

a long way." At the same time, however, crimes against animals go on, reducing their numbers daily. Some believe that the media has been slow to respond.

REVERSING THE TIDE

New efforts to save rare species are being made every day. Computers and satellites help to track down smugglers and poachers. Radio collars are used to keep track of animals in the wild. Zoos are breeding animals to prevent their extinction. This is called "breeding in captivity." The Oklahoma City Zoo, for example, is trying to save the black rhinoceros. This rhino is almost extinct. Poachers have cut its numbers from 65,000 to 3,500 in the past 20 years.

Our national parks are better managed now. And campers are more aware. They know how snowmobiles and boats hurt wildlife and how chemicals kill creatures of the sea. They know that when trees are cut down, other species can die. "Education is the best way to save our wildlife," a ranger in Yellowstone has said.

Concerned groups are more active now, too. The Audubon Society gives rewards for reporting killers of bears in Yellowstone. Through the work of the World Wildlife Fund (WWF), 99 countries have agreed to limit the import and export of 40,000

Thanks to zoo breeding programs, endangered species like the black rhino
are beginning to grow again in numbers.

plants and animals. The WWF helped to bring about the worldwide ban on the trade of ivory. Its Elephant Action Campaign hopes to put poachers out of business.

In 1979 Greenpeace was formed. This organization promotes the rights of wildlife. To do so, it takes direct, nonviolent action. For example, Greenpeace has placed its boats in the way of those trying to kill whales. Many whales have been saved by its action.

In 1989 Japan ordered a ban on ivory. Until then, Japan had bought 40 percent of the ivory in the world. One month later, CITES declared a worldwide ban on ivory. And the elephant was officially placed on the endangered species list.

Many Americans have joined French explorer Jacques Cousteau in his campaign to turn Antarctica into the first world park. By doing so, oil drilling—and oil spills—would end there. The whole area would become a wilderness where all wildlife could live in peace.

Californians were upset when President George Bush supported an oil-drilling plan for California's coast. Environmentalists worked hard to oppose it. President Bush asked top scientists to study the plan, and on November 3, 1989, they gave their report: "No more drilling," they said. This was a great victory for all those who care about endangered species. A final decision on the drilling has not been made.

Because of protests by environmental groups like Greenpeace, several countries have passed laws restricting the hunting of whales.

SIGNS OF HOPE

Today there are signs of hope in the wild and in the zoos. In 1987 twelve red wolves were placed in North Carolina's coastal wild. In 1988 two wolf pups were seen. The number of Chesapeake Bay bald eagles has grown from 121 to 161. Bald eagles are also nesting in Quabbin Reservoir in Massachusetts. A second population of California sea otters is thriving off the coast of Los Angeles. In Africa zebras and other rare species are kept safe on preserves. It is hoped that one day they, too, can increase their numbers and go back to the wild.

American scientists are developing a new ecosystem in a devastated forest in Peru—one of the world's last great unspoiled wilderness areas. "Miracles still happen," one worker said.

THE FOOD CHAIN

"Everything that lives is a part of the food chain," said biologist Paul Ehrlich. All beings are interdependent. The best way to help save our world's species is to learn to recognize how humans break the chain that links us all. Often people do not know when they are harming wildlife. They do not understand how species are related. They forget that when they damage the world habitat, they damage themselves, too. There are probably 51 million species on our planet. Humans form just one of them. So it helps to know how the food-chain process works.

Bald eagles, once on the endangered species list, have now been reclassified as "threatened." Breeding programs and strict laws protecting them have allowed their numbers to increase.

Frogs, for example, feed on insects. When polluted ponds kill frogs, insects spread over the area in harmful numbers. In Africa rats were not a problem when the big cats were there to eat them. But the death of the cats meant the increase of the rats. Rats spread diseases that can kill other species—including humans.

HOW CAN I HELP?

There are many ways children and adults alike can help save wildlife. Knowing wildlife laws will help. The U.S. Fish and Wildlife Service will send a copy of *Facts About Federal Wildlife Laws* to keep you up-to-date. When you know the laws, you can report violations to your local game warden. Joining the World Wildlife Fund will help, too. WWF offers toys, books, T-shirts, and other gifts to its members. Profits from all items are used to help save species around the world.

Government action is very important in protecting habitats. You can write to your representative and senator to let them know about your concern. For example, you might want to support Jacques Cousteau's effort to protect the Antarctic. Or you might want to support the scientists' call to stop offshore oil drilling. Maybe you want to express your concern about logging or burning rain forests.

If no one buys ivory products, there will be no reason to kill elephants.

Zoos in some cities offer "Adopt an Animal" programs. Donations, ranging from $10 to $2,000, help to feed an animal for one year in the zoo. Adopting an animal in your name helps to bring animals and humans closer together. Zoos also offer helpful facts about local wildlife and habitats. A visit to a wildlife refuge helps to build respect for animals in their own habitats.

In Texas conservationist Tony Amos started beach cleanup groups. He found out that poachers were not the only killers of wildlife in the world. Innocent sunbathers were harming animals and birds every day! Each time a plastic six-pack ring is tossed onto the beach, a turtle or bird is put in danger. The rings could strangle them to death. Beach patrols need everyone's help. Tony began with just a few helpers, young and old. Today 6,000 people arrive every year to clean 120 miles of beach! "Countless birds and other species have been saved," he said. Twenty-five states have such programs.

Most of us are not able to go to Africa to see the results of poaching and smuggling. But we can join the Elephant Action Campaign and other efforts of the World Wildlife Fund. Ten years ago there were 1.3 million elephants in Africa. Today there are 625,000, and one is being killed every ten minutes. The ban on ivory trade may slow down the killing. But wildlife experts agree that we can help by knowing—and not buying— the products for which the animals are killed.

Today the number of endangered species approaches 900. In the coming years, more than a thousand will be added to the list. Endangered plant species exceed 200, according to the U.S. Fish and Wildlife Service. The process of listing a species is not simple. It requires careful research and public hearings. Only 60 a year

Plastic six-pack rings can be deadly to birds, which become tangled in them. Some communities have formed beach patrols to pick up litter on their shores.

Game preserves provide an opportunity for many species to live in a protected environment.

make the list. And we continue to lose a species a day. Biologist Paul Ehrlich says, "Rates of extinction are far beyond any in history."

GOOD-BYE, KALAHARI

After seven years in the African wilderness, Delia and Mark Owens left the Kalahari Desert and returned to the United States. In 1985 their book *Cry of the Kalahari* won the John Burroughs Medal for the best natural history book of the year. Delia and Mark had achieved their goal—to do the best they could to help save wildlife, not only in Africa but throughout the world. Their kinship with wildlife is best described in Mark's own words:

> Most of the animals we found there had never seen humans before. They had never been shot at, chased by trucks, trapped, or snared. On a rainy morning we would often wake up with 3,000 antelope grazing around our tent. Lions, leopards, and brown hyenas visited our camp at night and woke us up by tugging the tent ropes. Sometimes they sat in the moonlight with us, and they even smelled our faces.

Humans may be wildlife's most feared predator, but they can be its savior, too.

With help, endangered species like the California gray whale may once again thrive.

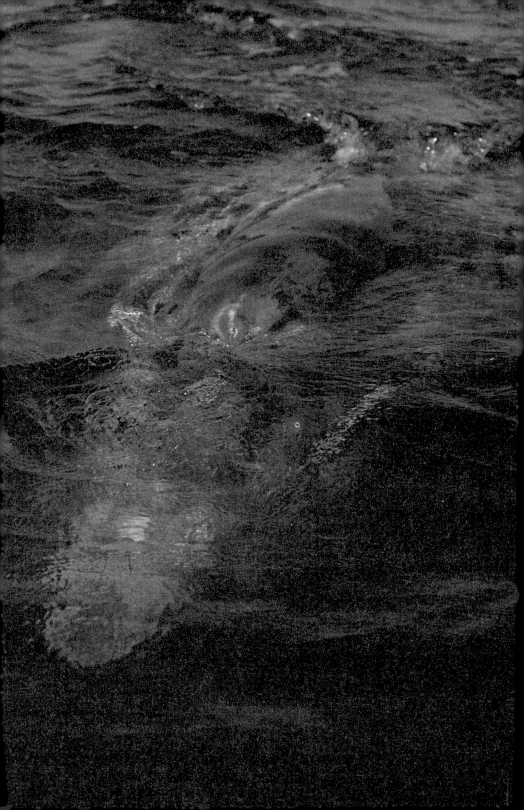

The beautiful Bengal tiger is one of the many species endangered by human carelessness.

FOR MORE INFORMATION

Department of the Interior
U.S. Fish and Wildlife Service
18th and C Streets NW
Washington, DC 20240

Greenpeace
1611 Connecticut Avenue NW
Washington, DC 20009

National Audubon Society
950 Third Avenue
New York, NY 10022
(212) 832-3200

National Wildlife Federation
1412 16th Street NW
Washington, DC 20036-2266
(202) 797-6800

The Nature Conservancy
Suite 800
1800 N. Kent Street
Arlington, VA 22209
(703) 841-5300

Sierra Club
730 Polk Street
San Francisco, CA 94109
(415) 981-8634

The Wilderness Society
1400 I Street NW, 10th Fl.
Washington, DC 20005
(202) 842-3400

World Wildlife Fund
1250 24th Street NW
Washington, DC 20037
(202) 293-4800

FOR FURTHER READING

Benyus, Janine M. *The Field Guide to Wildlife Habitats of the Western United States.* New York: Simon & Schuster, 1989.

Burt, Olive. *Rescued! America's Endangered Wildlife on the Comeback Trail.* Englewood Cliffs, NJ: Julian Messner, 1980.

Cox, James A. *The Endangered Ones.* New York: Crown, 1975.

Felder, Deborah G. *The Kids' World Almanac of Animals and Pets.* New York: World Almanac, 1989.

Laycock, George. *World Endangered Wildlife.* New York: Grosset & Dunlap, 1973.

Miller, Susanne Santoro. *Whales and Sharks and Other Creatures of the Sea.* New York: Simon & Schuster, 1982.

Penny, Malcolm. *Endangered Animals.* New York: Franklin Watts, 1988.

Whitaker, John O., Jr. *The Audubon Society Field Guide to North American Mammals.* New York: Alfred A. Knopf, 1988.

Whitcombe, Bobbie. *A Big Book of Animals.* Newmarket, England: Brimax Books, 1989.

GLOSSARY

biologist *One who is schooled in the science of living forms, their growth and evolution.*

conservation *The protection of natural resources such as forests, waterways, and wildlife habitats.*

ecologist *One who studies living forms in relation to their surroundings.*

ecosystem *A community of living forms together with their surroundings (environment).*

endangered *In danger of becoming extinct.*

environmentalist *One who tries to protect the environment.*

evolution *A gradual process in which things change into new forms, usually better ones.*

extinct *No longer in existence.*

habitat *The area where a species normally lives and relies upon for food.*

interdependent *Mutually dependent; dependent on one another; in need of each other.*

mutation *The process of being altered or changed.*

naturalist *One who is schooled in natural history (the study of natural life and evolution).*

poacher *An illegal hunter/killer of endangered species and other wildlife.*

predator *An animal that lives by preying on (hunting, catching, eating) others.*

ranger *One who patrols and protects forests and helps campers and tourists.*

smuggler *One who brings in or takes out animals or products illegally.*

species *A distinct group of animals or plants that have common qualities.*

zoology *The science of animals.*

INDEX